AF885078

Cats In Space
Félicette: The Cosmic Cat

When you think of animals in space, you probably picture a dog or a monkey, so hold onto your helmets! Meet Félicette, a brave little black and white "Tuxedo" cat, who went from wandering the streets of Paris, France, all the way to outer space.

Fast Facts from the Field

- **From Streets to Stars:** Félicette started as a stray cat in Paris. Scientists picked her because she was healthy and had the best temperament over 13 others in training.
- **Rocket Rider:** In 1963, she zoomed 100 miles up in a rocket called Véronique. That is higher than 17.8 Mount Everests stacked on top of each other!
- **Zero-G Floating:** When the rocket reached the very top, Félicette got to float in zero gravity for 5-7 minutes. Scientists wanted to study how gravity changes would affect cats' brains.
- **The "Cat-sule" Landing:** Her spaceship separated from the rocket and used parachutes to slow down. She landed safely in the Saharan desert 13 minutes after blasting off.
- **A Statue for a Star:** People forgot her for a long time (or called her "Felix" by mistake!), but today she has her very own bronze statue to honor her amazing adventure.

SUMI DUSK

Made in United States
Cleveland, OH
04 November 2026

47647243R00069